I'm FOUR Times Tabler!

Bill Gillham and Mark Burgess

A Magnet Book

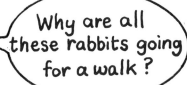

Why are all these rabbits going for a walk?

Why are they all in fours?

$$1 \times 4 = 4$$

$2 \times 4 = 8$

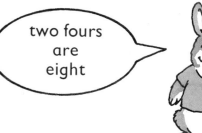

$3 \times 4 = 12$

three fours are twelve

$4 \times 4 = 16$

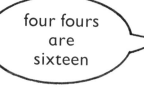

$$5 \times 4 = 20$$

$$6 \times 4 = 24$$

six fours
are
twenty-four

$$7 \times 4 = 28$$

$8 \times 4 = 32$

$$9 \times 4 = 36$$

$$10 \times 4 = 40$$

11 × 4 = 44

eleven fours are forty-four

$$12 \times 4 = 48$$

$1 \times 4 = 4$

$2 \times 4 = 8$

$3 \times 4 = 12$

$4 \times 4 = 16$

$5 \times 4 = 20$

$6 \times 4 = 24$

$7 \times 4 = 28$

$8 \times 4 = 32$

$9 \times 4 = 36$

$10 \times 4 = 40$

$11 \times 4 = 44$

$12 \times 4 = 48$

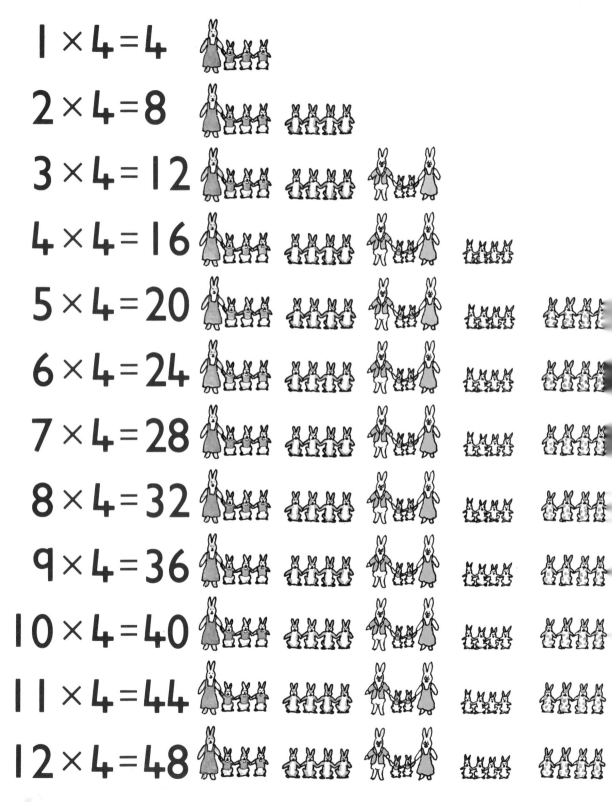

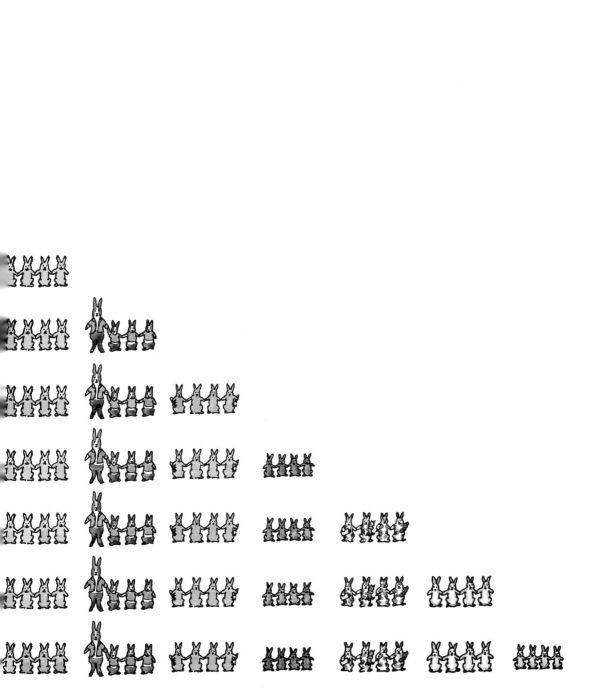

Activities

BUTTON STENCILS

Choose a *large* coat button which is easy to handle
and has four-pattern sewing holes big enough
to get a pencil-tip through.

Show the child how to draw round the button and
then make a four-dot pattern by putting the
pencil-tip firmly through the holes.

The stencils should be made in one line and
the child encouraged to recite the table
along the line, 'One four is four . . .'
and so on.

FOUR-FOLD
CUT-OUTS

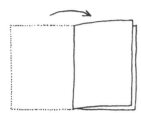

Fold an A4 sheet in half. Then fold each flap
back to the centre, like this.

Press it together and draw a rabbit on the front,
making sure the arms reach the side of the paper.

Cut carefully along the lines you have drawn,
so that when opened out the four rabbits
are linked together.

Make twelve rabbit paperchains and ask the child
to draw in the eyes and nose for each rabbit as
you do them. Each 'family' can be coloured differently
then set out in a row so that the child can recite
the table along the line.

NUMBER SQUARE PATTERN

Draw a 10 by 5 number square (the bigger the better) in black felt-tip pen:

1	2	3	4	5	6	7	8	9	10
11	12	13	14	15	16	17	18	19	20
21	22	23	24	25	26	27	28	29	30
31	32	33	34	35	36	37	38	39	40
41	42	43	44	45	46	47	48	49	50

Ask the child to colour in every fourth number
(using a coloured pencil rather than felt-tip).
Point out how it makes a pattern, then ask him
to count in fours pointing to the appropriate number
as he does so: 4 – 8 – 12 etc.
Explain that when you say the four times table
you are also *counting* in fours.

PLASTICINE DOGS

With the child make 12 plasticine dogs – bodies and heads but no legs. Vary the colours as far as possible.

Cut 24 matches in half – easier to manage and less wobbly than full-length matches. Ask the child to sort out *four* legs for each dog, before fixing them in to make the dogs stand up.

Then ask the child to say the table along the line as each dog is put in place, finishing off by counting in fours. When the child is confident in reciting the table, divide the dogs into two groups (or three or four) and ask, 'How many dogs are there?' and, 'How many legs altogether?'

Children need to know their tables because:
– simple multiplication, *which you can do in your head,*
is a skill of practical use in everyday life;
– the number patterns and groupings that occur
in tables help them to understand more advanced
mathematical concepts like *sets,* number *series*
and *progressions.*

The Times Table Books teach these ideas in a clear
and enjoyable fashion and show vividly what happens
when you multiply.

Dr Bill Gillham is senior lecturer in the
Department of Psychology at Strathclyde University.

First published in Great Britain in 1987
as a Magnet original
by Methuen Children's Books Ltd
11 New Fetter Lane, London EC4P 4EE
Text copyright © 1987 Bill Gillham
Illustrations copyright © 1987 Mark Burgess
Printed in Great Britain

ISBN 0 416 00222 6